Embracing the Green Revolution

Sustainable Agriculture Strategies For Food Production And Consumption

Hazel Rivera

Table of Contents

Table of Contents

1 Introduction to Sustainable Agriculture

Sustainable agriculture isn't just a buzzword; it's a necessity. It's about creating a food system that meets the needs of the present without compromising the ability of future generations to meet their own. This means considering the economic viability of farming practices alongside their environmental and social impacts. We can't simply grow food; we must grow it responsibly, ensuring the health of our planet and the well-being of those who depend on it. The challenge is significant, demanding innovation and a fundamental shift in how we approach food production.

The global food system is incredibly complex. Billions rely on it daily, yet it faces immense pressure. Population growth necessitates increased food production, but conventional agricultural methods often strain resources, leading to environmental degradation. Climate change further complicates matters, impacting crop yields and increasing the frequency of extreme weather events. A sustainable system must be resilient, adaptable, and capable of feeding a growing population under increasingly challenging circumstances. Successfully navigating these hurdles requires a multifaceted approach involving technological advancements, policy changes, and a shift in consumer behavior.

Defining sustainability within agriculture demands a nuanced understanding. It's more than just environmental protection; it encompasses economic viability and social equity. Sustainable agricultural practices must be profitable for farmers, ensuring their livelihoods are secure while safeguarding the environment. Furthermore, they must contribute to fair and just food systems, ensuring access to nutritious food for all, regardless of socioeconomic status. This means considering the entire chain, from farm to table, minimizing waste, and prioritizing equitable distribution. A truly sustainable approach necessitates a holistic view, integrating these three pillars—environmental, economic, and social—into a unified framework.

Consider the impact of soil degradation. Intensive farming practices, such as monoculture and excessive tillage, deplete soil nutrients and compromise its structure, leading to reduced fertility and increased erosion. This negatively affects crop yields, requiring increased inputs like fertilizers and pesticides, ultimately creating a vicious cycle. Sustainable practices prioritize soil health, employing methods such as cover cropping, crop rotation, and no-till farming. These techniques improve soil structure, enhance nutrient retention, and reduce the need for synthetic inputs. The resulting healthier soil leads to more resilient crops, contributing to both higher yields and a reduced environmental footprint. This is not just about improving the soil; it is about creating a foundation for long-term food security.

Water scarcity is another major challenge facing agriculture. Inefficient irrigation methods can waste vast quantities of water, exacerbating drought conditions in many regions. Sustainable water management involves implementing efficient irrigation techniques, such as drip irrigation and rainwater harvesting. These methods minimize water waste while delivering water directly to plant roots, improving water use efficiency. Selecting drought-resistant crops is another crucial strategy, reducing the overall water demand of agricultural production. Integrating water conservation practices into farming operations is not only environmentally responsible, it is a sound economic strategy, securing resources for future use.

The role of integrated pest management (IPM) cannot be overstated. Reliance on synthetic pesticides harms beneficial insects, contaminates water sources, and poses risks to human health. IPM emphasizes a preventative approach, utilizing a combination of methods to control pests while minimizing the use of pesticides. This might include biological control, using natural predators to control pest populations, cultural controls such as crop rotation, and employing pesticides only as a last resort. The long-term benefits of IPM extend beyond environmental protection; by reducing pesticide use, farmers can save money and produce safer, healthier food. The implementation of IPM strategies requires a shift in mindset, emphasizing prevention over immediate reaction. However, it is undeniably a more sustainable and ethical path forward in food production.

Sustainable agriculture is not simply a set of practices; it's a paradigm shift, requiring collaboration, innovation, and a long-term perspective. It demands a commitment to responsible resource management, equitable social structures, and resilient food systems. This chapter serves as a foundation, providing essential insight into the fundamental principles underpinning the transition towards a more sustainable and secure future for our food systems. It lays the groundwork for further exploration of the specific techniques and strategies that contribute to this crucial transition, setting the stage for a comprehensive understanding of the green revolution and its impact on food production and consumption.

1.1 Defining Sustainability

Sustainability isn't a single, easily defined concept; it's a multifaceted ideal encompassing environmental, economic, and social well-being. It's not merely about protecting the environment; it's about creating a system that thrives across all three dimensions, ensuring a healthy planet and a just society for present and future generations. This interwoven nature necessitates a holistic approach, rejecting simplistic solutions in favor of complex strategies. We need to understand that economic prosperity cannot come at the cost of environmental degradation or social inequality.

Consider the impact of industrial agriculture. While providing food for billions, its reliance on synthetic fertilizers, pesticides, and fossil fuels carries significant environmental costs, contributing to climate change, biodiversity loss, and soil degradation. Economically, this model can be unsustainable in the long term, vulnerable to price fluctuations and reliant on external inputs. Socially, it can displace communities and create exploitative labor conditions, underscoring the interconnectedness of sustainability's components.

A truly sustainable agricultural system, therefore, needs to move beyond maximizing short-term yields. It must prioritize long-term resilience, fostering a robust and adaptable food system capable of withstanding shocks such as climate change and disease outbreaks. This necessitates a shift towards integrated and diversified farming practices, mimicking the complexity and resilience of natural ecosystems. Sustainable agriculture aims not just to feed the population but to do so in a way that benefits the environment, supports fair labor practices, and ensures economic viability for farmers.

This isn't just about implementing individual practices, such as using less water or reducing pesticide use, though these are crucial. It also requires a significant shift in mindset, valuing long-term benefits over short-term gains, and recognizing the interconnectedness of our food system. We must move away from viewing agriculture as an isolated sector and instead understand its intricate connections to other parts of society and the environment. Economic models need to reflect the true costs of food production, accounting for environmental externalities like pollution and greenhouse gas emissions.

Therefore, defining sustainability in agriculture requires going beyond simplistic definitions. It necessitates a comprehensive understanding of the complex interactions between environmental, economic, and social factors. It involves rethinking our relationship with food production, moving beyond a focus solely on efficiency and yield towards a model that values biodiversity, resilience, and equity. We must recognize that sustainable agriculture isn't just a set of practices but a paradigm shift, requiring collaboration across sectors, innovative technologies, and a fundamental change in societal values and consumption patterns.

The path towards true sustainability involves a systemic approach, one that acknowledges that environmental damage, economic instability, and social injustice are intertwined challenges that must be addressed holistically. It requires collaboration among farmers, scientists, policymakers, and consumers to create a food system that is both productive and environmentally responsible, socially

equitable, and economically viable for generations to come. Only then can we build a future where food security and environmental sustainability are not opposing goals but mutually reinforcing elements of a thriving global society. This requires a profound reevaluation of our current practices and a commitment to innovative solutions, technological advancements, and mindful consumption choices.

This integrated approach acknowledges the limitations of solely focusing on a single aspect of sustainability. For instance, while organic farming may reduce environmental impact, it may not always be economically viable for all farmers. Similarly, technological advancements, like precision agriculture, can improve resource use efficiency, but they require upfront investment and may exacerbate inequalities if not implemented equitably. Therefore, crafting a successful path towards sustainability necessitates considering the unique context of each region and the specific challenges and opportunities facing local food systems.

In conclusion, defining sustainability in agriculture transcends simply ticking boxes; it involves a profound reassessment of our current food production systems and the creation of resilient, equitable, and environmentally sound alternatives. It demands innovative solutions, a holistic perspective, and collective action by all stakeholders to ensure the long-term well-being of both people and the planet. This commitment, reflected in our agricultural practices and consumption habits, is crucial for securing a sustainable future for food systems worldwide. The challenge lies not just in adopting sustainable practices, but in fostering a cultural shift that values long-term ecological and social well-being above short-term gains. Only through such a fundamental shift in perspective can we hope to truly embrace the green revolution and ensure a secure and sustainable food future for all.

1.2 Global Food Systems

Global food systems are incredibly complex, interwoven networks encompassing production, processing, distribution, consumption, and waste management. Consider the sheer scale: billions of people rely on these systems daily for sustenance, and these systems themselves are shaped by diverse economic, social, and environmental factors that constantly interact. Understanding these dynamics is crucial for building a more sustainable and equitable future. The challenges are monumental – feeding a burgeoning global population while safeguarding the planet's resources requires a paradigm shift.

We need to acknowledge the inequalities inherent in current food systems. Many farmers, particularly in developing nations, face significant challenges in accessing resources, technology, and markets. This often leads to low yields, economic hardship, and unsustainable practices. Addressing these disparities requires policy interventions that promote fair trade, empower smallholder farmers, and invest in rural development. These interventions must be tailored to the specific contexts of diverse communities, ensuring their voices are heard and their needs are met. Failure to address these issues will perpetuate inequities and hinder the transition to more sustainable food systems.

Technological advancements offer promising solutions. Precision agriculture, using technologies like GPS and sensors, allows farmers to optimize resource use, target pesticide applications, and improve yields. Similarly, advancements in crop breeding are producing more resilient varieties better adapted to changing climates and resistant to pests and diseases. However, these technologies are not universally accessible, raising concerns about equity and potential exacerbations of existing inequalities if not implemented thoughtfully. Accessibility, affordability, and effective training are crucial to ensure their beneficial impact on all. Sustainable intensification methods, which increase output while minimizing environmental impact, are crucial to maximizing agricultural production in the face of growing demand.

Beyond technological innovation, we must also tackle the issue of food waste. A significant portion of food produced globally is lost or wasted during production, processing, distribution, and consumption. This waste represents a considerable loss of resources—water, energy, land—and contributes to greenhouse gas emissions. Reducing food waste requires a multi-pronged approach involving improved infrastructure, better storage facilities, more efficient transportation systems, and shifts in consumer behavior. Educational campaigns aimed at raising awareness of the problem and providing practical tips on reducing waste at home are equally necessary.

Ultimately, creating truly sustainable global food systems necessitates a holistic approach. This means integrating ecological considerations, social justice, and economic viability. It requires collaboration among governments, businesses, researchers, and communities. Effective policies are indispensable, creating incentives for sustainable practices and discouraging unsustainable ones. International cooperation is also vital for knowledge sharing, technology transfer, and coordinated action on issues like climate change, which significantly impacts food production. Building resilience into food systems, by diversifying production and promoting robust supply chains, will ensure food security even in the face of unexpected shocks. The goal is not merely to feed the world, but to do so equitably and sustainably, protecting both people and the planet. The challenges are significant, but the potential rewards—a more secure, just, and environmentally sound future—are immense. Failure to address these profound challenges will result in an unstable future, burdened by food insecurity and environmental damage. We must choose a path toward lasting, positive change.

2 Sustainable Farming Practices

Crop rotation is a cornerstone of sustainable farming. It's not just about planting different crops in sequence; it's a strategic approach to soil health, pest management, and overall farm productivity. By rotating crops, you disrupt the life cycle of pests and diseases, reducing the need for harmful pesticides. Legumes, for instance, fix nitrogen in the soil, naturally fertilizing the land for subsequent crops. This reduces reliance on synthetic fertilizers, lowering production costs and minimizing the environmental impact associated with their manufacturing and transport. Understanding the specific needs of different plants—their nutrient requirements, growth patterns, and susceptibility to pests—is crucial for effective crop rotation. Careful planning ensures that crops with contrasting needs follow each other, creating a balanced and resilient agricultural system. For example, a heavy feeder like corn could be followed by a legume to replenish soil nitrogen, then a light feeder like wheat to allow the soil to recover. This cyclical approach creates a synergistic effect, improving soil structure and fertility over time, leading to healthier plants and higher yields. Moreover, the varied root systems of different crops improve soil aeration and water infiltration, further enhancing soil health.

Integrated Pest Management (IPM) offers a more holistic and environmentally friendly approach to pest control compared to relying solely on chemical pesticides. IPM involves a multi-pronged strategy that combines various techniques to minimize pest populations while protecting beneficial insects and the environment. This approach prioritizes preventative measures, such as choosing pest-resistant crop varieties and maintaining healthy soil conditions. Regular monitoring of pest populations is crucial to identify problems early on, allowing for timely intervention before significant damage occurs. Biological control, using natural enemies like beneficial insects or nematodes, can be a powerful tool in IPM. Other methods such as using physical barriers, crop rotation, and sanitation help to prevent pest infestations, thus reducing the need for chemical intervention. When chemical intervention is deemed necessary, IPM emphasizes the use of the least toxic pesticides possible and only at the lowest effective dose. Proper application techniques minimize the risk to non-target organisms and the environment. The adoption of IPM requires careful planning, knowledge of local pest dynamics, and the willingness to adapt strategies as conditions change. However, the long-term benefits—reduced environmental damage, lower pesticide costs, and enhanced farm profitability—make IPM a vital component of sustainable farming. Organic farming practices shun synthetic inputs, prioritizing natural methods to maintain soil fertility, control pests, and manage crops. It's a system built around ecological principles, emphasizing biodiversity, soil health, and minimizing environmental impact. Organic certification involves adhering to strict guidelines, ensuring the integrity of the process. Organic farmers rely on cover crops to improve soil structure and fertility, suppressing weeds, and preventing erosion. Composting and manure application provide natural sources of nutrients, enriching the soil while reducing the reliance on synthetic fertilizers. Pest control methods primarily use biological controls, such as introducing natural predators or employing beneficial insects. Rotating crops helps to break pest cycles and enhance soil health. The rejection of synthetic pesticides and fertilizers significantly reduces the risk of water contamination and enhances biodiversity within and around the farm. While organic farming may have a slightly lower yield in certain situations compared to conventional agriculture, the long-term benefits, including improved environmental stewardship and increased soil health, make it a sustainable and increasingly important approach to food production. The higher market value of organic produce often compensates for lower yields, making it an economically viable option for many farmers. Choosing the right organic farming techniques requires careful consideration of the local climate, soil type, and pest pressure; thus, organic farming is a complex but rewarding approach to sustainable agriculture.

2.1 Crop Rotation Techniques

Crop rotation isn't just about switching crops yearly; it's a strategic dance between plants, soil, and the environment. Think of it as a carefully orchestrated ballet, where each plant plays a unique role in nourishing the soil and keeping it vibrant and productive. Successful crop rotation involves understanding the specific needs of various plants and how they interact with the soil ecosystem. A diverse selection of crops contributes to a healthy and balanced soil microbiome. For instance, legumes like beans and alfalfa enrich the soil with nitrogen, while other crops, like corn or wheat, benefit from this nitrogen boost. This natural fertilization reduces or eliminates the need for synthetic nitrogen fertilizers, minimizing environmental impact.

Choosing the right crops to rotate is crucial and depends heavily on several factors. Your local climate plays a significant role: a crop suited for a hot, dry climate might not thrive in a cool, wet one. Consider your soil type. Sandy soils require different crop rotations than clay soils. The length of your growing season also dictates the selection. Longer seasons allow for more diverse rotations, while shorter seasons might necessitate simpler ones. Economic factors, such as market demand and crop prices, must also be considered in creating a sustainable and profitable rotation.

Effective crop rotation helps minimize pest and disease infestations. Planting a variety of crops disrupts the life cycles of many pests and pathogens, hindering their ability to establish a strong population. It's a natural pest management strategy that reduces or even eliminates the need for harmful pesticides. For example, rotating a susceptible crop like potatoes with a resistant crop like oats can significantly reduce the spread of potato blight. This approach is healthier for both the environment and consumer health. In addition, the diversity of roots from different crops improves soil structure, allowing for better water infiltration and drainage.

Beyond pest and disease control, crop rotation contributes significantly to improved soil fertility. Legumes, as previously mentioned, are nitrogen-fixing plants, enriching the soil naturally. Other crops, like those with deep root systems, help to improve soil structure by breaking up compacted soil and increasing aeration. The decomposition of crop residues also enhances soil organic matter, leading to improved water retention and nutrient cycling. This organic matter not only feeds the soil but also supports a rich biodiversity of beneficial organisms, further enhancing soil health and plant vigor.

Planning your crop rotation isn't simply about following a predetermined scheme; it's an ongoing process of learning and adaptation. Begin by carefully observing your land. Pay close attention to the growth and health of each crop, noting any signs of nutrient deficiencies or pest infestations. Keep detailed records of your crop yields, noting any variations across different crops and years. This data will allow you to fine-tune your rotation system and make informed decisions about future planting. Consider the long-term health of your soil as a primary goal, remembering that healthy soil leads to healthy crops.

Remember, crop rotation is a long-term investment, yielding benefits over several years. It's not a quick fix for poor soil health but rather a holistic approach to sustainable agriculture. It enhances environmental sustainability by reducing reliance on synthetic fertilizers and pesticides and improving water management. It also contributes to the economic sustainability of your farm through increased yields and reduced input costs. Furthermore, you'll be contributing to a healthier planet by adopting a natural and environmentally friendly approach to food production. The reward for thoughtful planning and diligent execution is enhanced soil health, increased yields, and long-term farm sustainability. Start small, learn from your experiences, and gradually refine your rotation strategy. The journey toward sustainable agriculture is a continuous process of learning and adaptation; embrace the challenge and enjoy the fruits of your labor.

2.2 Integrated Pest Management

Integrated pest management (IPM) is a holistic approach to pest control that prioritizes preventing pest problems and minimizing the use of pesticides. It's not about eliminating pests entirely—that's often impossible and environmentally damaging—but rather about keeping pest populations below economically damaging levels. This involves a thorough understanding of the pest's life cycle, its natural enemies, and the ecosystem in which it thrives. Effective IPM requires careful monitoring and targeted interventions, maximizing the use of environmentally friendly methods first.

Successful IPM relies heavily on careful monitoring. Regular inspections of crops allow for early detection of pests, enabling swift and targeted intervention before populations explode and cause significant damage. Knowing what to look for is crucial. This involves understanding the signs and symptoms of pest infestations, learning to identify various pests and their life stages, and recognizing the damage they cause to plants. Accurate identification helps in determining the best management strategy. Simple tools such as sticky traps, visual inspections, and pheromone traps are often sufficient.

Once pests are detected, the next step is to choose the least harmful control method. This is the heart of IPM. Cultural controls, which involve adjusting farming practices, are often the first line of defense. These techniques include crop rotation, proper planting dates, healthy soil management, and the use of pest-resistant varieties. These methods aim to create an environment that is less favorable for pests to thrive. For example, choosing a crop variety that is inherently less susceptible to a particular pest dramatically reduces the need for pesticides.

Biological control offers a more active, yet still environmentally sound, approach. This involves using natural enemies of the pests—predators, parasites, or pathogens—to suppress their populations. Ladybugs devouring aphids, parasitic wasps controlling caterpillars, and naturally occurring bacteria killing harmful insects are all examples of this approach. Introducing these beneficial organisms into a farm's ecosystem can significantly reduce the reliance on chemical pesticides and create a more balanced agricultural environment. The understanding of the ecological dynamics involved is crucial for maximizing their effectiveness.

Only if cultural and biological controls prove insufficient should chemical pesticides be considered, and even then, their application must be highly targeted and precise. Choosing the right pesticide is critical. This includes considering its toxicity to non-target organisms, including beneficial insects and pollinators, and the potential for environmental contamination through runoff or drift. Proper application techniques are essential to minimize adverse effects on the environment. This means using only the recommended amount of pesticide, at the right time, and employing appropriate application methods to avoid unnecessary exposure.

IPM requires a long-term perspective. It's not a quick fix but rather a continuous process of monitoring, assessment, and adjustment. The effectiveness of different management strategies varies depending on factors like the specific pest, the crop being grown, environmental conditions, and regional variations. Maintaining detailed records of pest activity, control measures employed, and their effectiveness is vital for adapting IPM strategies over time. The data gathered informs future decisions, making the process adaptive and increasingly efficient. Regular evaluation and modification are key components to success.

Beyond the farm itself, IPM requires collaboration and knowledge sharing. Farmers can learn from each other's experiences, share data on pest outbreaks, and collectively develop effective regional strategies. Extension services and research institutions play a crucial role in providing farmers with the information and training necessary to implement IPM successfully. This collaborative approach builds resilience within the agricultural community, and strengthens overall pest management practices. This shared approach benefits everyone.

The benefits of IPM extend beyond simply reducing pesticide use. It can also lead to improved crop yields by preventing significant damage and ensuring healthier plants. The reduced reliance on synthetic pesticides can improve water quality, protect beneficial insects like pollinators, and help maintain biodiversity within agricultural ecosystems. IPM contributes to sustainable agriculture by minimizing environmental impacts and improving the long-term health of the farming system. Considering both short and long-term impacts is critical for implementing this system successfully. Implementing IPM requires a commitment to continuous learning and adaptation. Farmers must stay informed about new pests, emerging control techniques, and best management practices. This ongoing educational process helps ensure that IPM strategies remain effective and adaptable to changing conditions. The dynamic nature of pest populations requires continual attention, monitoring, and adjustments to the chosen practices. This constant learning and adaption is critical for long-term success.

Successfully integrating IPM into farming practices leads to several significant benefits. It offers environmental protection by minimizing the use of harmful chemicals, safeguards the health of beneficial organisms that contribute to a balanced ecosystem, and improves the overall sustainability of food production systems. By reducing reliance on synthetic pesticides, IPM contributes to food security by enhancing the health and productivity of farming systems. The long-term sustainability offered by this approach benefits both individual farmers and the entire agricultural ecosystem.

2.3 Organic Farming Methods

Organic farming prioritizes ecological balance and soil health, rejecting synthetic pesticides, fertilizers, and genetically modified organisms (GMOs). This approach fosters biodiversity, improving both the soil's ecosystem and the resilience of the crops. The long-term benefits extend beyond the immediate harvest, creating a sustainable agricultural system capable of withstanding environmental changes and market fluctuations. It's a holistic approach demanding a deeper understanding of natural processes.

Successful organic farming demands meticulous planning and observation. Farmers must become experts in their local ecosystems, identifying beneficial insects and understanding the complexities of soil biology. This intimate knowledge allows for precise interventions, minimizing risks and maximizing yields. They need to monitor soil health rigorously, adapting their practices according to the specific needs of the land. Essentially, they're fostering a living partnership with nature.

Pest and disease management is crucial, and organic farmers employ a range of strategies. This could involve crop rotation, the introduction of beneficial insects, companion planting, and the use of organic pesticides derived from natural sources. These strategies often require more time and effort than conventional methods, but they reduce the environmental impact significantly, avoiding the harmful effects of synthetic pesticides on human health and the environment. Integrated pest management, or IPM, is a core element.

Organic fertilization relies heavily on natural materials, such as compost, manure, and cover crops. These enhance soil fertility naturally, improving its structure and water retention capacity. The process of creating compost demands patience and attention to detail, but it rewards the farmer with nutrient-rich soil amendments. This method reduces the reliance on synthetic fertilizers that can deplete soil nutrients, contaminate groundwater, and contribute to greenhouse gas emissions.

Weed management presents unique challenges, often demanding more labor-intensive practices like hand-weeding or mechanical methods. These techniques help preserve soil health and reduce the need for herbicides, which can harm beneficial organisms and potentially contaminate waterways. Mulching and cover cropping also play a vital role in weed suppression, creating a healthier environment for the plants. Each method requires careful consideration based on the specific context of the farm.

Crop rotation is fundamental to maintaining soil fertility and reducing pest and disease pressure. By systematically changing the crops grown in a particular field each year, farmers disrupt the life cycle of pests and replenish soil nutrients. This is a low-input approach that improves long-term productivity and enhances the soil's natural capacity to support plant life. The selection of rotation crops should carefully balance nutritional needs and pest-resistance capabilities.

The transition to organic farming often requires a significant investment of time and effort. Farmers need to understand the specific regulations for certification, learn effective organic farming techniques, and adapt their operations to meet the demands of the organic market. This means navigating new challenges, dealing with potential initial yield reductions, and managing the economic aspects of changing farming methods.

Marketing organic produce effectively is a key aspect of success. Building relationships with local buyers, farmers markets, and co-operative enterprises can help ensure consistent demand. This requires an effective marketing strategy that highlights the unique qualities of organically grown products – their superior flavor, nutritional value, and minimal environmental impact. It is a strategy centered on sustainability and ethical production.

Despite the challenges, the rewards of organic farming are numerous. It promotes environmentally sound practices, enhances biodiversity, safeguards human health, and provides a more sustainable pathway for future food production. It also strengthens rural communities, fostering resilience and promoting long-term ecological stability. It's not simply a farming method; it's a philosophy that embraces the interconnectedness of the environment and human well-being.

Continuous learning and adaptation are essential for organic farmers. Staying informed on the latest research, best practices, and emerging challenges is crucial for ongoing success. Participation in workshops, conferences, and farmer networks allows for the exchange of knowledge and experiences, fostering innovation and improving resilience in the face of ever-changing conditions. It is a journey of constant refinement and discovery, fueled by a passion for sustainable food production.

3 Water Management in Agriculture

Water scarcity is a significant challenge for modern agriculture. Farmers often rely on inefficient irrigation methods, leading to substantial water waste and unsustainable practices. This results not only in environmental damage but also economic losses for farmers struggling with rising water costs and diminished yields. A shift toward efficient irrigation systems is crucial. Drip irrigation, for instance, delivers water directly to plant roots, minimizing evaporation and runoff, thereby maximizing water use efficiency. Furthermore, advancements in sensor technology enable precise monitoring of soil moisture levels, allowing for targeted irrigation only when and where needed. Implementing these technologies can significantly reduce water consumption while maintaining or even improving crop yields.

Precise water management requires a comprehensive understanding of the local climate and soil conditions. Different crops have varying water requirements, and understanding these differences is crucial for optimizing irrigation schedules. Farmers should consult with agricultural extension services or utilize readily available soil moisture sensors to gain a precise understanding of their particular needs. Integrating weather forecasting into irrigation planning enables proactive adjustments to account for anticipated rainfall and minimize unnecessary water use. This approach minimizes water waste while ensuring sufficient hydration for optimal plant growth.

Beyond technological solutions, water conservation strategies play a vital role in sustainable water management. Selecting drought-resistant crop varieties is a simple yet effective method to reduce water demand. These varieties are genetically engineered or naturally adapted to thrive in arid or semi-arid conditions, requiring significantly less irrigation. Additionally, the practice of rainwater harvesting—collecting rainwater for later use—can provide a supplementary water source, decreasing reliance on external water supplies. This approach contributes not only to water conservation but also reduces reliance on energy-intensive pumping systems.

Improved water management is about more than just technology and conservation. The adoption of these practices involves a significant educational component. Farmers need comprehensive training on proper irrigation techniques, water-saving technologies, and efficient water management strategies. Government agencies and non-governmental organizations play a crucial role in providing this training, coupled with accessible resources like online guides and workshops. This investment in agricultural education translates to improved practices across the farming community, leading to a more sustainable and resilient agricultural sector.

Water management also influences the overall health of the agricultural ecosystem. Over-irrigation can lead to soil salinity, harming plant health and reducing yields. Conversely, inadequate irrigation can stress crops, leaving them vulnerable to pests and diseases. Finding the right balance is critical to maintain both healthy plants and soil, essential to sustainable agriculture. Careful monitoring and adjustments, incorporating data from soil sensors and weather forecasts, are key to navigating this delicate equilibrium. This requires careful attention and responsive adjustments to changing conditions.

Furthermore, effective water management is intrinsically linked to economic considerations. While the initial investment in new technologies or infrastructure may seem substantial, the long-term cost savings from reduced water usage and increased crop yields often outweigh the upfront expenses. Reduced water bills, decreased reliance on expensive external water supplies, and improved crop yields contribute to improved farm profitability, fostering the long-term sustainability of agricultural practices. This underscores the importance of considering the holistic economic implications of adopting sustainable water management practices. A comprehensive cost-benefit analysis will often reveal significant long-term advantages.

Lastly, successful water management requires collaboration and cooperation. Farmers, policymakers, researchers, and other stakeholders must work together to develop and implement effective water management strategies. This collaborative approach allows for the sharing of best practices, the development of tailored solutions, and the establishment of supportive policies that promote responsible water use across the agricultural sector. Sharing knowledge and coordinating efforts ensures that the limited resources are utilized in the most responsible and sustainable way possible, protecting both the environment and the long-term viability of agriculture. This cooperative model is critical to address the water challenges that face the future of agriculture.

3.1 Efficient Irrigation

Efficient irrigation is crucial for sustainable agriculture. It's not just about watering crops; it's about optimizing water use to maximize yields while minimizing environmental impact and resource depletion. This means employing methods that deliver the right amount of water at the right time, directly to the plant's root zone, reducing evaporation and runoff. A poorly designed irrigation system can lead to waterlogging, nutrient leaching, and increased salinity in the soil, harming plant health and reducing productivity.

Conversely, a well-designed system is an investment in long-term profitability and environmental stewardship.

Consider drip irrigation, a technique that delivers water directly to the roots through a network of tubes and emitters. This method minimizes water waste by reducing evaporation and runoff. Drip irrigation is particularly effective for orchards, vineyards, and high-value crops, where precise water control is essential for optimal fruit development and quality. Studies have shown that drip irrigation can significantly increase yields and reduce water consumption compared to traditional methods like furrow irrigation. Moreover, it reduces weed growth by keeping the soil surface dry, lowering the need for herbicides.

Micro-sprinklers offer another efficient irrigation approach, providing a gentle spray of water over a small area. This technique is suitable for a wider range of crops than drip irrigation, including vegetables and field crops. Micro-sprinklers are especially beneficial in areas with sandy soils, where water infiltration is high. Unlike flood irrigation, which often leads to uneven water distribution and significant water loss, micro-sprinklers provide consistent and targeted watering. This translates into improved water use efficiency and reduced labor costs compared to manual watering.

Choosing the right irrigation system depends on many factors. These factors include the type of crop, soil conditions, water availability, and budget considerations. For example, farmers in arid regions may opt for efficient drip or micro-sprinkler systems, while those in areas with abundant water might use sprinkler irrigation, though perhaps with improved technology to reduce waste. Analyzing your specific situation is key. Consider consulting with an agricultural engineer or irrigation specialist to assess your needs and choose the most appropriate system. They can provide tailored recommendations for system design, water management practices, and maintenance routines.

Proper irrigation management extends beyond choosing the right technology. Regularly monitoring soil moisture levels is crucial for making informed decisions about when and how much water to apply. Soil moisture sensors are readily available and provide real-time data, allowing farmers to respond quickly to changing conditions. Avoid overwatering, which leads to waterlogging and nutrient leaching, and underwatering, which can stress crops and reduce yields. Regularly inspecting irrigation systems is also vital to identify and address leaks promptly. Even small leaks can waste considerable amounts of water over time and increase your operational costs.

Efficient irrigation also involves integrating water conservation strategies. Practices like rainwater harvesting, where rainwater is collected and stored for later use, can significantly reduce reliance on external water sources, reducing your environmental impact and water bills. Implementing drought-resistant crops can also contribute to efficient water management. Selecting crop varieties adapted to drier conditions minimizes water stress and improves water use efficiency. The use of mulch around plants can further reduce evaporation and conserve soil moisture. It also helps to prevent weed growth.

Beyond the technology and practices themselves, education and training are paramount. Farmers, agricultural workers, and students should understand the principles of water management in agriculture, including the different irrigation techniques, their advantages, and limitations. This includes understanding the importance of regular monitoring, maintenance, and the economic and environmental benefits of efficient irrigation. There are numerous resources available, including workshops, online courses, and extension services, which can provide essential knowledge and practical skills. This knowledge empowers farmers to make informed decisions about water management, leading to improved yields, reduced costs, and a more sustainable future for agriculture.

3.2 Water Conservation

Water scarcity is a significant threat to global food security, particularly in arid and semi-arid regions. Efficient water management is no longer a luxury; it's a necessity for sustainable agriculture. This chapter delves into practical strategies for conserving water in agricultural practices, focusing on approaches that are both effective and readily implementable.

The cornerstone of water conservation in agriculture lies in the choice of irrigation techniques. Flood irrigation, while seemingly simple, often leads to significant water loss through evaporation and runoff. Drip irrigation, on the other hand, delivers water directly to the plant roots, minimizing waste. Similarly, sprinkler irrigation, while less efficient than drip, is still preferable to flood irrigation in many situations, particularly where topography is challenging. Farmers should carefully assess their specific circumstances—soil type, terrain, crop needs—to determine the most efficient irrigation method for their operation. The selection should also incorporate the available resources and budget. A well-designed irrigation system, incorporating water sensors and automated control, can significantly reduce water consumption while optimizing crop yield.

Beyond irrigation methods, selecting appropriate crop varieties plays a crucial role. Drought-resistant crops, engineered to tolerate water stress, are invaluable in regions with limited water resources. These crops, often developed through traditional breeding techniques or genetic modification, exhibit resilience to dry conditions, minimizing the need for supplemental irrigation. Moreover, integrating drought-resistant varieties into crop rotation strategies can further improve water use efficiency while diversifying the farm's produce and enhancing overall soil health. This is especially important in areas experiencing increasing drought intensity and frequency due to climate change. Research on drought-tolerant crops is ongoing, with promising advances in genetic engineering, offering farmers a wider selection of resilient options in the near future.

Water harvesting techniques, including rainwater harvesting and the construction of small reservoirs, offer farmers a measure of control over water availability. Rainwater harvesting involves collecting and storing rainwater for later use, supplementing irrigation needs during dry periods. The design and implementation of rainwater harvesting systems will vary widely depending on local rainfall patterns, topography, and soil type. However, many simple, cost-effective systems exist that farmers can easily adopt with appropriate training and support. Similarly, constructing small reservoirs or ponds can provide a readily available water source for irrigation. These systems can also enhance biodiversity by creating habitats for beneficial insects and other organisms.

Soil moisture monitoring is increasingly crucial for effective water management. Simple tools such as soil moisture probes or advanced sensors integrated into irrigation systems provide real-time information on soil water content, allowing farmers to precisely adjust irrigation schedules. This reduces the risk of over-watering, minimizing water waste while ensuring optimal plant growth. The adoption of such technologies requires initial investment and training; however, the long-term benefits in terms of water and resource savings often outweigh the initial costs.

Beyond these technological and management strategies, effective water conservation also requires a shift in agricultural practices. No-till farming, for example, reduces water evaporation from the soil surface by maintaining soil cover. Similarly, cover cropping helps retain moisture in the soil, reducing the need for irrigation. Incorporating these practices into overall farm management helps to build resilience in the face of water scarcity, simultaneously fostering enhanced soil health and productivity. The transition to these practices, however, may necessitate changes to existing farming equipment and techniques, and ongoing technical support will be essential for widespread adoption.

Furthermore, promoting water-efficient agricultural practices requires a multifaceted approach. Government policies that incentivize water conservation, such as subsidies for efficient irrigation technologies or water-saving crop varieties, are critical. Educational programs that equip farmers with the knowledge and skills to adopt these techniques are equally essential. Through collaboration between governments, research institutions, and farmer organizations, effective strategies for water conservation in agriculture can be developed, implemented, and continuously improved to ensure long-term food security and sustainability. This necessitates a concerted effort to address both technical and socioeconomic aspects of efficient water management in the agricultural sector. Only a collective response, encompassing effective policies, technical innovation, and farmer empowerment, will ensure sustainable water use in agriculture.

4 Soil Health and Fertility

Healthy soil is the cornerstone of productive and sustainable agriculture. Its structure, rich in organic matter and teeming with beneficial microorganisms, dictates everything from nutrient availability to water retention. Ignoring soil health is akin to building a house on a cracked foundation; eventual collapse is inevitable. Investing in soil health is not just an environmental imperative, it's a sound economic strategy.

Improving soil structure is crucial. Compacted soil restricts root growth, limiting a plant's access to water and nutrients. Practices like no-till farming, where seeds are sown directly into the undisturbed soil, minimize compaction. Cover crops, planted between cash crops, further improve soil structure by preventing erosion and adding organic matter. Rotating crops also enhances structure; different plants have diverse root systems that help break up compaction and improve aeration. Consider incorporating diverse cover crop mixes to further improve the soil's overall health. You might use legumes for nitrogen fixation, grasses for erosion control, and brassicas for pest control, creating a natural synergy.

Nutrient management is more than just applying fertilizers; it's about optimizing nutrient cycles. Excessive fertilizer use leads to nutrient runoff, polluting waterways and harming ecosystems. Soil testing helps determine the actual nutrient needs of your crops, minimizing unnecessary inputs. Composting and using manure are excellent ways to return nutrients to the soil naturally, significantly reducing reliance on synthetic fertilizers. These methods are less costly in the long run and contribute to a healthier soil environment. Remember to compost your kitchen scraps and yard waste; this valuable material improves soil quality and reduces waste. Integrating animal manure adds organic matter and enhances soil fertility. Careful planning ensures your nutrient management plan aligns with your specific needs and the local environmental conditions.

Soil erosion, the gradual wearing away of topsoil, is a severe threat to agricultural productivity. Wind and water carry away fertile soil, reducing yields and causing sedimentation in waterways. Contour plowing, terracing, and cover cropping are effective erosion control measures. Contour plowing follows the natural contours of the land, slowing water runoff. Terracing creates level platforms on slopes, reducing erosion by breaking up steep gradients. Cover crops act as a protective shield, reducing the impact of wind and rain on the soil surface. By strategically using these techniques, you protect your investment in the long run, maintaining fertile land that ensures yields. Careful land management practices, such as maintaining buffer strips and incorporating appropriate tree cover, can significantly slow soil erosion.

Cover cropping offers numerous benefits beyond erosion control. These plants enrich the soil with organic matter, improving its structure and water-holding capacity. Some cover crops, like legumes, fix atmospheric nitrogen, reducing the need for nitrogen fertilizers. Cover crops also suppress weeds, reducing competition for nutrients and water. Integrating cover cropping into a sound rotation plan, considering the species' impact on various crops, adds a layer of sophistication to your sustainable agricultural strategy. A diversity of cover crops contributes to a resilient and robust system.

The effective implementation of these soil health practices requires careful planning and observation. Regular soil testing provides invaluable data for informed decision-making. Furthermore, mindful observation of your crops and soil conditions can help identify potential problems early on, allowing for timely intervention. Sustainable agriculture isn't a one-size-fits-all approach, so adapting these strategies to suit your local climate, soil type and crop choices is vital. Continuously monitor and adapt your strategies over time to achieve optimal soil health and productivity. Remember this is a long-term commitment – improving soil health is not instantaneous but pays significant dividends over the years.

4.1 Improving Soil Structure

Soil structure, the arrangement of soil particles into aggregates, is paramount to healthy, productive land. A well-structured soil possesses a balanced porosity, allowing for optimal water infiltration, aeration, and root penetration. Conversely, poor soil structure, often characterized by compaction or excessive erosion, hinders these essential processes, leading to reduced crop yields and environmental degradation. Improving soil structure is, therefore, a cornerstone of sustainable agriculture, demanding a multifaceted approach.

The most straightforward method involves minimizing soil disturbance. Conventional tillage practices, while effective in short-term weed control, can disrupt soil aggregates, leading to compaction and loss of organic matter. No-till farming, on the other hand, leaves crop residues on the soil surface, shielding it from erosion and fostering the growth of beneficial soil organisms that bind soil particles together. This protective layer also helps regulate soil temperature and moisture, creating a more stable environment for plant growth. Adopting no-till methods necessitates careful consideration of weed management and nutrient cycling, but the long-term benefits for soil structure overwhelmingly outweigh these challenges.

Cover cropping offers another powerful tool for improving soil structure. Planting cover crops during fallow periods adds organic matter to the soil, providing food for soil organisms. These organisms, in turn, create stable soil aggregates, increasing water infiltration and drainage. Different cover crops have varying effects; some excel at improving soil structure, others at suppressing weeds, and still others at fixing nitrogen. A well-designed cover cropping program should incorporate a diverse mix of species to maximize these benefits, creating a dynamic soil ecosystem robust against various stressors. The selection of appropriate cover crops depends on local climate, soil type, and the subsequent main crop. Organic matter plays an absolutely crucial role; it acts as a glue, binding soil particles together to form stable aggregates. Increasing organic matter content is, therefore, essential for improving soil structure. This can be achieved through several methods, including the addition of compost, manure, or other organic residues. Compost, a product of controlled decomposition of organic materials, is particularly rich in beneficial microorganisms and humus, a stable form of organic matter that enhances soil structure. Manure, while offering similar benefits, requires careful management to prevent nutrient runoff and potential pathogen introduction. The application of these organic amendments not only improves soil structure but also enhances nutrient availability and water retention capacity. Careful consideration must be given to the specific type of organic matter added and the rate of application to avoid negative consequences.

Beyond organic matter addition, other practices contribute to improved soil structure. Maintaining appropriate soil pH is crucial; many soil organisms thrive within a specific pH range. Extreme pH values can damage soil structure, hindering the activity of beneficial organisms. Regular soil testing can help identify pH imbalances and guide appropriate corrective measures. Furthermore, careful management of irrigation and drainage can prevent compaction and waterlogging, both of which negatively impact soil structure. Effective irrigation techniques such as drip irrigation deliver water directly to plant roots, minimizing soil disturbance. Well-designed drainage systems prevent waterlogging, reducing soil compaction. These techniques contribute to a more stable soil structure, fostering a healthier root system and ultimately improving crop yields.

The improvement of soil structure isn't a one-time event; it's an ongoing process requiring consistent effort and careful management. It's a journey of nurturing a living soil, fostering its biodiversity and resilience. By implementing these techniques, farmers can create a more productive and environmentally sustainable agricultural system. Remember, healthy soil underpins healthy crops and a healthy environment, so every effort spent on soil health yields multiple rewards. The long-term benefits – increased crop yields, reduced water and fertilizer use, and enhanced carbon sequestration – clearly demonstrate that investment in soil structure is an investment in the future of sustainable agriculture. The methods outlined above, though diverse, are interconnected, and their synergistic effect is far greater than the sum of their individual contributions.

4.2 Nutrient Management

Nutrient management is the cornerstone of sustainable agriculture, impacting yield, soil health, and environmental impact. Effective nutrient management hinges on understanding soil nutrient levels, crop needs, and minimizing environmental damage from excess fertilizers. Soil testing is paramount; it provides a precise picture of your soil's nutrient composition, revealing deficiencies and excesses, guiding fertilizer application to prevent overuse and pollution. This data-driven approach, contrasted with blanket application methods, offers significant economic advantages and significantly reduces environmental harm.

Choosing the right fertilizer is equally crucial. Synthetic fertilizers offer a readily available source of nutrients, but overuse leads to nutrient runoff, polluting waterways and harming ecosystems. Organic fertilizers, like compost and manure, provide nutrients gradually, improving soil structure and enhancing microbial activity, offering a more environmentally friendly, though often more labor-intensive, alternative. The selection should consider both cost and environmental factors, a balance achievable through careful planning and appropriate soil testing.

Timing is also essential. Nutrients should be applied when the crop can efficiently utilize them, maximizing uptake and minimizing losses. This means avoiding applications during heavy rainfall or periods of high evaporation, which can lead to nutrient leaching and atmospheric emissions of nitrogen. Careful timing can significantly improve nutrient use efficiency and reduce environmental damage. Consider splitting fertilizer applications to provide nutrients throughout the growing season, maximizing plant growth while optimizing nutrient uptake.

Precision agriculture technologies offer even more fine-tuned nutrient management. GPS-guided fertilizer spreaders and variable rate technology allow for precise application based on soil variations within a field. This ensures that nutrients are only applied where needed, reducing waste and environmental impact, while optimizing nutrient use efficiency and yields. Investment in such technologies might seem substantial upfront, but the long-term benefits of reduced fertilizer usage and improved yields justify the initial cost.

Minimizing nutrient runoff is of crucial importance. Utilizing cover crops between cash crops is an effective strategy. Cover crops prevent soil erosion and improve soil structure, thereby retaining nutrients and preventing them from leaching into water bodies. They also enhance soil health by improving its organic matter, enhancing nutrient availability to cash crops. Different cover crops offer various benefits; some excel at nutrient uptake while others help improve soil structure.

Beyond fertilization, crop rotation plays a crucial role in nutrient management. Rotating crops with varying nutrient demands helps maintain soil fertility. Legumes, for example, fix nitrogen in the soil, reducing reliance on nitrogen fertilizers in subsequent crops. Careful planning of crop rotations allows for improved nutrient cycling, reducing the need for external inputs and optimizing the soil's natural resources.

Integrating livestock grazing into crop production can improve nutrient cycling. Animal manure provides a natural fertilizer, enriching the soil with nutrients. Strategic grazing management can further enhance nutrient use efficiency and reduce the need for synthetic fertilizers. Careful consideration of grazing intensity and timing is essential to prevent overgrazing and soil degradation.

Monitoring nutrient levels and adjusting strategies over time is key to continuous improvement. Regular soil testing and yield monitoring provides crucial feedback, enabling adjustments to fertilizer applications and other management practices. This iterative process allows for refining nutrient management strategies, optimizing nutrient use efficiency, and minimizing environmental impact. ensuring long-term sustainability and financial efficiency. In conclusion, successful nutrient management requires a holistic approach, encompassing soil testing, fertilizer selection, timing, precision agriculture technologies, cover cropping, crop rotation, livestock integration, and continuous monitoring and adaptation. This meticulous approach is indispensable to sustainable agriculture and to securing the future of food production.

4.3 Soil Erosion Control

Soil erosion, the gradual wearing away of topsoil, poses a significant threat to agricultural productivity and environmental health. It diminishes soil fertility, reduces water-holding capacity, and increases the risk of flooding and sedimentation. Ignoring soil erosion is akin to depleting a precious natural resource, jeopardizing the very foundation of food production for generations to come. Understanding the mechanisms behind soil erosion and implementing effective control measures is paramount for sustainable agriculture. Effective soil erosion control hinges upon a multi-pronged approach. One key strategy involves the careful management of land use. This means avoiding overgrazing, which compacts the soil and removes protective vegetation. It also involves selecting appropriate cropping systems, minimizing bare soil exposure during fallow periods, and employing techniques like contour farming, which follows the natural contours of the land to reduce water runoff. These practices protect the soil from the erosive forces of wind and water. Terracing, a more structural approach, reshapes sloped land into a series of level platforms. These platforms create a series of small, manageable areas, reducing the speed and volume of water flow. This significantly minimizes soil erosion on steep slopes, where the threat is most severe. Proper terrace construction requires expertise and careful planning to prevent collapse and ensure long-term effectiveness. Improperly built terraces can actually exacerbate erosion problems.

Cover cropping is another powerful tool. Planting vegetation, especially legumes, during fallow periods significantly reduces soil erosion. These plants protect the soil from the elements, reducing wind and water erosion. The plant roots also improve soil structure, enhancing water infiltration and reducing runoff. Cover crops also add organic matter to the soil, boosting its fertility and creating a healthier ecosystem. Choosing the appropriate cover crop for the specific environmental conditions is crucial for maximizing its benefits.

Windbreaks, strategically planted rows of trees or shrubs, create a barrier against wind erosion, reducing the force of the wind and trapping soil particles. These are especially effective in arid and semi-arid regions where wind erosion is a significant concern. The selection of windbreak species needs to consider local climate and soil conditions to ensure the windbreaks thrive and remain effective over time. Regular maintenance, including pruning and replacement of damaged plants, is also essential to maintain their protective function.

Conservation tillage practices, like no-till farming, minimize soil disturbance, leaving crop residues on the surface to protect the soil. This reduces erosion, improves soil structure and water infiltration, and enhances soil organic matter content. However, choosing the correct equipment and management strategies for no-till farming is important, as improper implementation can lead to weed problems and other challenges.

In addition to these physical measures, sustainable agricultural practices play a vital role. Crop rotation, the practice of systematically changing crop types in a field, helps maintain soil health, reduces pest and disease pressure, and increases the soil's ability to resist erosion. This strategy reduces the dependence on fertilizers and pesticides, which often contribute to soil degradation.

Finally, effective communication and education are critical. Farmers must be equipped with the knowledge and resources to implement these erosion control measures. Government programs, educational initiatives, and technical assistance from experts can greatly improve adoption rates and increase the long-term success of erosion control efforts. Sharing best practices and showcasing successful examples through farmer-to-farmer learning can also accelerate the implementation of effective soil erosion control strategies across agricultural landscapes. Sustainable soil management requires a concerted effort from individuals, communities, and governmental organizations. Investing in soil erosion control isn't just about protecting farmland; it's about securing a sustainable future for agriculture and the environment.

4.4 Composting and Manure

Composting and manure are cornerstones of sustainable agriculture, offering a powerful combination to enhance soil health and reduce reliance on synthetic fertilizers. Composting, the natural breakdown of organic matter, creates a rich, nutrient-dense amendment teeming with beneficial microbes. This process, while seemingly simple, involves a careful balance of materials – browns (like dried leaves and twigs) and greens (such as grass clippings and fruit scraps) – to ensure proper aeration and decomposition. The resulting compost improves soil structure, water retention, and nutrient availability, leading to healthier plants and increased yields. Successful composting requires understanding the decomposition process, regularly turning the pile to maintain oxygen levels and managing moisture to avoid anaerobic conditions that produce unpleasant odors.

Manure, the waste from livestock, is another valuable resource for enriching agricultural land. Its composition varies considerably depending on the animal source and its management. For example, poultry manure is high in nitrogen, while cow manure is richer in phosphorus. Proper manure management, including appropriate storage and application techniques, is crucial to avoid nutrient runoff and potential pollution of waterways. Understanding the nutrient content of different manures enables farmers to tailor applications to specific crop needs, optimizing fertilizer use and reducing the environmental footprint of agriculture. Moreover, applying manure directly reduces reliance on synthetic fertilizers, cutting down on production costs and minimizing the environmental impacts associated with their manufacturing and transportation.

Successfully integrating composting and manure into your farming system requires careful planning and management. For example, consider the scale of your operation. A small-scale farm may effectively manage composting and manure applications using simple methods, while large-scale operations may require more sophisticated approaches involving specialized machinery and storage facilities. Proper storage is crucial for preventing nutrient loss and environmental contamination, particularly with liquid manures. Effective storage methods reduce the risk of leaching, which can lead to water pollution and nutrient losses.

Further complicating this equation are regulatory frameworks, which vary by region and often stipulate restrictions on manure application to prevent water contamination. Knowledge of these regulations is paramount. For instance, certain nitrogen limits might necessitate precise calculations regarding manure application rates. It is important to comply with all applicable environmental regulations, seeking advice from local agricultural extension offices or environmental agencies to ensure responsible and legal practices. Also, precise record-keeping is vital for demonstrating compliance with regulations, particularly for larger-scale farming enterprises. Beyond the regulatory aspects, understanding the different types of composting and manure management systems is crucial. These range from simple in-situ composting directly within the soil to sophisticated systems employing large-scale composting facilities and advanced manure management technologies. The choice of system is heavily dependent upon factors like the scale of operation, the types of crops cultivated, and resource availability. In addition to traditional methods, innovative technologies are continually being developed to improve composting and manure management efficiency. For instance, some systems leverage aerobic digestion to accelerate the decomposition process and enhance nutrient recovery.

Moreover, effective manure and compost utilization should always consider soil testing. Regular soil analysis helps farmers determine nutrient levels and adjust manure and compost applications accordingly. This approach prevents excessive nutrient application, thereby minimizing pollution risk and maximizing the efficiency of these organic inputs. Soil testing also helps to ensure the soil is well balanced and receiving the optimal amounts of necessary nutrients for successful crop production. This scientific approach to resource management emphasizes a holistic understanding of soil properties, ensuring that soil conditions remain optimal and that nutrients are utilized efficiently.

In essence, effective utilization of composting and manure presents a path towards truly sustainable agriculture. By carefully considering the various aspects of each system, including storage, regulatory compliance, technological advancements, and soil testing, farmers can dramatically improve soil health, reduce their environmental impact, and promote long-term farm productivity and profitability. The successful integration of these approaches represents a fundamental shift toward more environmentally sound and economically sustainable agricultural practices. In closing, remember that commitment to continuous learning and adaptive management is paramount for maximizing the benefits of composting and manure in your farming system. The knowledge gained through these sustainable soil management practices can be pivotal to the successful cultivation of sustainable agriculture for future generations.

4.5 Cover Cropping

Cover cropping, a deceptively simple practice, offers profound benefits to soil health and overall farm sustainability. It involves planting specific crops, not for direct harvest, but to improve soil conditions. These "cover crops" can be legumes like clover or alfalfa, grasses like rye or oats, or even brassicas like mustard. The choice depends largely on your specific climate, soil type, and agricultural goals. Effective implementation requires careful consideration of several factors, making it a more complex strategy than initially perceived.

Unlike other soil improvement methods, cover cropping adds a layer of ecological complexity, fostering biodiversity above and below the ground. The diverse root systems of different cover crops improve soil structure, creating channels for better water infiltration and aeration. This improved structure reduces compaction, a common problem in intensively farmed lands, leading to healthier roots for subsequent cash crops and enhanced water retention capabilities within the soil profile. The increased porosity and permeability benefit both plant growth and overall soil health.

Furthermore, cover crops excel at suppressing weeds. Their dense growth shades the soil surface, preventing weed seeds from germinating. Their competitive root systems deprive weeds of essential nutrients and water, limiting weed proliferation. This reduces the need for herbicides, saving farmers money and contributing to a more environmentally sound agricultural practice. By reducing weed competition, cover crops increase the efficiency of water and nutrient use by cash crops, optimizing resource utilization and potentially increasing crop yield. This also reduces the risk of soil erosion, a common result of bare soil exposure and harsh weather conditions. Cover crops act as a living mulch, shielding the soil from the elements, and their residues increase soil organic matter, enriching the soil and enhancing nutrient cycling.

The benefits extend beyond weed suppression and soil structure. Many cover crops, particularly legumes, are nitrogen-fixing plants. These plants harbor symbiotic bacteria in their roots that convert atmospheric nitrogen into a form usable by plants. This natural fertilization reduces reliance on synthetic nitrogen fertilizers, which are expensive and contribute significantly to greenhouse gas emissions. By incorporating these cover crops into the soil through tillage or other methods, nitrogen is released for use by the following cash crop, cutting down on additional input costs for the farmer, reducing nitrogen runoff into waterways, thus decreasing the risks of water contamination.

Selecting the right cover crop mix is crucial for optimal results. Farmers must consider factors like climate, soil type, and the following cash crop when making their choices. For instance, a fast-growing crop like rye may be suitable for a short growing season to quickly protect the soil from erosion during winter months, while a legume like clover can be used to build soil fertility for a spring planting of vegetables or field crops. The planning and selection process should also consider the possibility of pest and disease suppression, or even potential allelopathic effects (chemical inhibition of growth of neighboring plants) that some cover crops may have on weeds and subsequent crops.

Proper management of cover crops is also critical for success. This includes planting at the appropriate time, using correct seeding rates, and effectively terminating the cover crop before planting the cash crop. Termination methods can range from physical means such as mowing or rolling to chemical methods, depending on the cover crop and the succeeding cash crop. However, it's important to choose termination methods carefully to minimize soil disturbance and maintain soil health. Furthermore, proper timing of termination is critical; it needs to be done before the cover crop begins to compete excessively with the main crop, ensuring that the beneficial effects of the cover crop are maximized without detrimental impacts on the yield of the cash crop.

In conclusion, cover cropping is not merely a soil improvement technique but a multifaceted sustainable agriculture practice. It significantly contributes to improving soil health, reducing reliance on synthetic inputs, and enhancing the overall environmental sustainability of farms. By carefully selecting and managing cover crops, farmers can reap numerous benefits, leading to healthier soils, increased crop yields, and a more environmentally responsible agricultural system. Cover cropping represents a practical and effective step towards a more resilient and sustainable agricultural future and should be widely promoted and adopted as a core practice within modern agricultural systems, leading to significant environmental and economic advantages for both farmers and consumers.

5 Reducing Agriculture's Carbon Footprint

Agriculture's contribution to climate change is undeniable. Methane emissions from livestock, nitrous oxide from fertilizers, and carbon dioxide from land use changes all contribute significantly to global warming. However, agriculture also presents a significant opportunity for carbon sequestration, a process by which carbon dioxide is removed from the atmosphere and stored in the soil and plants. This dual role – emitter and sequesterer – highlights the importance of adopting sustainable practices to minimize the former and maximize the latter.

Reducing emissions requires a multifaceted approach. Improving livestock management techniques, such as altering diets to reduce enteric fermentation, can significantly cut methane output. This involves transitioning to feedstuffs that produce less methane during digestion. Precision livestock farming technologies like sensors can also assist in monitoring livestock health and feed efficiency.

By identifying and addressing issues early, the overall efficiency and consequently reduced emissions of the livestock operation can improve. The increased efficiency reduces waste and maximizes resource use, minimizing the environmental impact.

Efficient fertilizer use is another critical area. Using precise application methods, such as variable-rate fertilization guided by soil testing and remote sensing, can greatly reduce nitrous oxide emissions. This precision application ensures fertilizers are applied in precise amounts and locations, maximizing efficiency, minimizing runoff and reducing associated pollution. Incorporating cover crops into cropping systems can also improve nitrogen use efficiency and reduce the need for synthetic nitrogen fertilizers. This strategy reduces the reliance on synthetic inputs while maintaining crop yields, thus minimizing environmental impact.

Beyond emissions reduction, carbon sequestration offers a powerful tool for climate change mitigation. No-till farming, which avoids soil disturbance, helps maintain soil organic carbon stocks. This practice protects the soil structure, improves water retention, and builds organic matter that acts as a carbon sink, storing atmospheric carbon. This, in turn, reduces the amount of carbon dioxide in the atmosphere, contributing to climate change mitigation. The addition of organic matter through cover cropping, composting, and the strategic application of manure further enhances carbon sequestration potential. These practices contribute to better soil structure, enhanced water infiltration, and improved nutrient availability.

Renewable energy sources are vital for reducing agriculture's reliance on fossil fuels. Solar panels can power irrigation pumps and other farm equipment, reducing greenhouse gas emissions associated with electricity generation. Similarly, biogas digesters can convert livestock manure into biogas, a renewable energy source that can be used for electricity generation or heating. Such renewable energy strategies lead to lower energy costs, reduced reliance on non-renewable sources and a lower carbon footprint. This diversified energy strategy also improves energy security and resilience of agricultural operations.

Furthermore, integrating agroforestry practices, like planting trees within or around agricultural lands, can enhance carbon sequestration and provide numerous other environmental and economic benefits. Trees not only sequester carbon but also provide shade for crops, reduce soil erosion, and improve biodiversity. This integrated approach shows the synergistic relationship between forestry and agriculture in mitigating climate change and enhancing ecological balance, leading to an environmentally friendly and economically stable system.

The implementation of these strategies requires a shift in mindset and investment in new technologies and practices. Farmers need access to training, financial incentives, and supportive policies to adopt these sustainable approaches. Government regulations and market incentives, such as carbon credits, can play a vital role in driving this transition. However, these strategies can be incredibly valuable, economically speaking, in lowering energy costs, and improving the health of the soil and consequently, the crops it produces, thus making the transition into a more sustainable system profitable. This financial stability ensures the widespread adoption of the environmentally friendly and profitable practices, leading to widespread change.

Ultimately, reducing agriculture's carbon footprint requires a holistic approach that integrates emissions reduction and carbon sequestration strategies. By embracing these practices, we can build a more sustainable and resilient agricultural sector, ensuring food security while mitigating climate change. This interconnectedness of the strategies highlights the necessity of a coordinated and multidisciplinary approach, involving farmers, scientists, policymakers, and consumers. This integrated, holistic approach towards the reduction of the carbon footprint of the agricultural sector ensures food security and climate change mitigation. The future of sustainable food production depends on its successful implementation.

5.1 Greenhouse Gas Emissions

Agricultural practices significantly contribute to greenhouse gas emissions, impacting global climate change. Methane, a potent greenhouse gas, is released during rice cultivation and livestock farming through enteric fermentation (digestion) in animals and manure management. Nitrous oxide, another powerful greenhouse gas, is emitted from the application of nitrogen fertilizers. Furthermore, the use of fossil fuels in agricultural machinery, transportation, and processing contributes substantially to carbon dioxide emissions. These emissions pose a significant challenge to achieving sustainable agriculture.

Reducing greenhouse gas emissions from agriculture requires a multi-pronged approach. One crucial step is optimizing fertilizer use. Precision agriculture techniques, such as variable rate application and sensor-based monitoring, can minimize nitrogen fertilizer application, thus lowering nitrous oxide emissions. Improving soil health through practices like cover cropping and no-till farming enhances nitrogen retention in the soil, reducing the need for synthetic fertilizers. Moreover, the incorporation of organic matter improves soil carbon sequestration, further mitigating climate change.

Livestock management practices also play a critical role. Improved feed efficiency reduces methane emissions per unit of animal product. Selecting breeds with lower methane emissions and managing manure efficiently, through anaerobic digestion for biogas production, can significantly decrease emissions. Dietary shifts towards plant-based protein sources can also contribute to lowering overall agricultural greenhouse gas emissions. The integration of these strategies, rather than isolated actions, is crucial for effective change.

Renewable energy sources can replace fossil fuels used in agriculture. Solar panels can power farm machinery and irrigation systems, reducing reliance on fossil fuels and minimizing carbon dioxide emissions. Biogas produced from animal manure can generate electricity and heat for farm operations, offering a sustainable alternative to conventional energy sources. Investing in renewable energy infrastructure on farms is not only environmentally beneficial but can also lead to cost savings in the long run.

Policy measures are crucial to incentivize the adoption of emission-reducing practices. Carbon pricing mechanisms, such as carbon taxes or emissions trading schemes, can make polluters pay for their environmental damage. Government subsidies can support the adoption of sustainable practices, such as precision agriculture and renewable energy technologies. Research and development funding can accelerate the innovation and deployment of emission-reducing technologies. Education and awareness campaigns are necessary to inform farmers about the importance of climate-smart agriculture and the available emission reduction options.

The quantification of greenhouse gas emissions from individual farms and across agricultural sectors is essential for effective management. Life cycle assessment (LCA) methodologies can be used to comprehensively evaluate the emissions associated with different agricultural products and practices. This data is invaluable for tracking progress towards emission reduction targets, identifying hotspots for emission reduction, and informing policy decisions. Ultimately, integrating these data-driven approaches into sustainable farming practices is key for transitioning to a low-carbon agricultural system.

Technological advancements hold tremendous potential for mitigating greenhouse gas emissions. Remote sensing technologies can monitor crop health and soil conditions, allowing for optimized fertilizer and water application, thereby reducing emissions. Precision livestock farming technologies can monitor animal health and behavior, enabling more efficient feed management and reducing methane emissions. Furthermore, the development and implementation of new technologies are crucial in the ongoing effort to achieve sustainable and emission-reducing agricultural processes.

The transition to a low-carbon agricultural system requires a concerted effort from farmers, policymakers, researchers, and consumers. While it involves challenges, the benefits of reducing greenhouse gas emissions are immense. Not only will it help mitigate climate change, but it will also enhance the long-term sustainability and resilience of agricultural systems. By adopting a holistic approach, encompassing technological innovation, policy support, and behavioral changes, we can move towards a future where agriculture contributes to environmental stewardship, rather than being a significant contributor to climate change. Embracing this holistic approach requires collaboration and a clear understanding of the interconnectedness of climate change, agriculture, and sustainable food production. The future of food security and environmental sustainability hinges on effective greenhouse gas emission mitigation strategies within the agricultural sector.

5.2 Carbon Sequestration

Carbon sequestration in agriculture offers a powerful tool for mitigating climate change while simultaneously enhancing soil health and crop productivity. It's not just about reducing emissions; it's about actively removing carbon dioxide from the atmosphere and storing it safely in the soil. This process involves several key strategies, each with its own set of benefits and considerations.

One of the most effective methods is improving soil organic carbon (SOC) levels. Healthy soils, rich in organic matter, act like giant carbon sponges, holding significant amounts of carbon. This organic matter comes from decomposing plant material, including roots, crop residues, and cover crops. The more organic matter added to the soil, the greater the carbon sequestration potential. Farmers can actively increase SOC by adopting practices like no-till farming, reducing soil disturbance, and incorporating cover crops into their rotations. These cover crops not only add organic matter, but also improve soil structure, reduce erosion, and suppress weeds. This holistic approach is crucial for long-term soil health and carbon storage.

Beyond simply adding organic matter, thoughtful management of grazing systems plays a significant role. Sustainable grazing practices, such as rotational grazing and managing livestock density, can significantly boost carbon sequestration in grasslands and pastures. These techniques mimic natural grazing patterns, preventing overgrazing and allowing for the growth of deep root systems that contribute to increased carbon storage in the soil. Properly managed grazing can enhance biodiversity, improve soil health, and offer economic benefits to farmers. However, overgrazing, conversely, can lead to soil degradation and release of stored carbon.

Another important consideration is the selection of appropriate crop species. Certain plants, like perennial grasses and deep-rooted legumes, are particularly effective at sequestering carbon due to their large biomass and extensive root systems. These plants also help improve soil structure and water retention, further contributing to overall soil health and carbon storage capacity. Careful consideration of crop rotation, incorporating species known for carbon sequestration, and employing diverse planting strategies are essential components for success. For example, integrating leguminous crops into rotations not only adds nitrogen to the soil, but also promotes the growth of deep root systems that efficiently sequester carbon. This symbiotic relationship between plants and soil microbes is crucial for enhancing the efficacy of carbon sequestration. Accurate measurement and monitoring of carbon sequestration is essential for evaluating the success of implemented strategies. This requires robust methods for assessing soil organic carbon content, employing techniques like soil sampling and laboratory analysis. Data collection and analysis provide crucial feedback for refining management practices and ensuring the long-term success of carbon sequestration initiatives. Regular monitoring also allows farmers to assess the environmental and economic benefits of their practices, making informed decisions about future strategies. This creates a cycle of improvement, where data informs practice, and practice leads to further data collection.

Finally, the potential of carbon sequestration in agriculture is immense, offering a vital contribution towards climate change mitigation. However, the success of these strategies depends on effective policy support, including incentives for farmers to adopt carbon-sequestering practices. Government programs and market mechanisms that reward farmers for carbon sequestration services can significantly incentivize adoption and help scale up these practices. This would support sustainable agricultural practices and facilitate the transition toward a more climate-resilient food system. Collaboration among scientists, policymakers, and farmers is essential to unlock the full potential of carbon sequestration and its contribution towards a sustainable and climate-resilient future.

Integrating these strategies effectively requires a long-term commitment, investment in research, and a holistic approach to land management.

5.3 Renewable Energy

Harnessing renewable energy sources is vital for minimizing agriculture's carbon footprint and ensuring the long-term viability of sustainable farming practices. Unlike fossil fuels, renewables offer a cleaner, more sustainable alternative, reducing greenhouse gas emissions and mitigating climate change, a critical aspect of true agricultural sustainability. The transition, however, demands a comprehensive understanding of available technologies and their effective implementation within agricultural settings.

Solar energy presents a readily accessible option for many agricultural operations. Photovoltaic (PV) systems can power irrigation pumps, lighting systems in greenhouses, and processing equipment, decreasing reliance on grid electricity generated from non-renewable sources. Furthermore, the decentralized nature of solar power makes it particularly suitable for farms located in remote areas with limited grid access, offering both energy independence and reduced transportation costs associated with fossil fuel delivery. Careful consideration must be given to the initial investment costs, system sizing based on energy demands, and maintenance requirements to ensure long-term cost-effectiveness and efficiency.

Wind energy, another promising renewable source, is particularly well-suited for larger farms or agricultural regions with consistent wind speeds. Wind turbines can generate substantial amounts of electricity, potentially exceeding a farm's energy needs and allowing for energy sales to the grid. However, the environmental impact of wind turbines, including potential effects on bird populations and landscape aesthetics, must be carefully assessed through environmental impact assessments before implementation. Moreover, initial capital investment and ongoing maintenance are substantial considerations that must be factored into the overall feasibility analysis. The ideal placement of wind turbines requires meticulous wind resource assessment to maximize energy generation.

Bioenergy, derived from agricultural residues and crops specifically grown for energy production, offers a circular approach to renewable energy within the agricultural sector. Biomass can be converted into biofuels like ethanol and biodiesel, replacing fossil fuel-based alternatives in farm machinery and transportation. The use of agricultural waste, such as crop stalks and manure, reduces waste disposal challenges while providing a sustainable energy source. This bioenergy production, however, should be carefully managed to avoid negative impacts on food production and land use, ensuring a balance between energy needs and food security. Moreover, the efficiency of biomass conversion technologies and the sustainability of feedstock cultivation must be rigorously evaluated.

Geothermal energy, while less widely applicable than solar or wind, can provide a reliable and constant source of energy in regions with geothermal resources. Geothermal energy can be harnessed for heating greenhouses, providing a stable temperature for optimal plant growth, and reducing reliance on fossil fuel-based heating systems. The initial investment for geothermal systems is significant, requiring careful evaluation of the available geothermal resource and the specific energy demands of the agricultural operation. It's essential to conduct thorough feasibility studies to justify the significant upfront costs and long-term benefits.

Integrating renewable energy into agricultural operations necessitates careful planning and consideration of various factors. Economic feasibility studies should be conducted to evaluate the cost-effectiveness of different renewable energy technologies based on energy needs, available resources, and long-term operational costs. Government incentives and funding opportunities for renewable energy projects should be actively explored. Furthermore, collaborations with energy experts and engineers are crucial for optimal system design, installation, and maintenance, ensuring long-term reliability and performance. The transition to renewable energy in agriculture is not merely a technological shift; it's a fundamental transformation toward a more sustainable and resilient food system.

6 Sustainable Food Consumption

Reducing food waste is paramount. We generate staggering amounts of edible food waste annually, contributing significantly to greenhouse gas emissions and resource depletion. Composting food scraps at home is a simple yet impactful change; diverting organic waste from landfills reduces methane production, a potent greenhouse gas. Furthermore, supporting initiatives aimed at rescuing and redistributing surplus food from supermarkets and restaurants reduces waste and ensures that nutritious food reaches those in need. Implementing better storage practices, from utilizing airtight containers to understanding "best before" dates, can extend the shelf life of produce and minimize spoilage. Careful meal planning, buying only what you need, and utilizing leftovers creatively are equally crucial for minimizing household waste. By consciously making these changes, you're directly contributing to a more sustainable food system.

Local and seasonal eating presents a wealth of environmental and economic advantages. Purchasing produce from local farmers' markets and farms prioritizes shorter transportation distances, reducing the carbon footprint associated with food miles. Seasonal produce, naturally grown during its optimal time, requires less energy for production, often necessitates fewer synthetic inputs, and often tastes superior. Moreover, supporting local farmers directly boosts the regional economy and strengthens community ties. The enhanced freshness of locally sourced food, its rich nutritional value compared to produce that has traveled long distances, and the overall positive impact on the local economy makes a strong case for integrating these practices into our daily lives. Embrace the diversity of seasonal ingredients—it's a culinary adventure and a sustainable choice.

Dietary choices exert a powerful influence on the environment. Shifting towards plant-based diets can drastically reduce your ecological footprint. Animal agriculture, particularly intensive livestock farming, is a significant contributor to greenhouse gas emissions, deforestation, and water pollution. Even reducing meat consumption, instead of eliminating it entirely, can make a considerable difference. Exploring alternative protein sources, such as lentils, beans, tofu, and tempeh, offers both nutritional and environmental benefits. Remember, it's not about perfection but about mindful choices that progressively lessen your impact. Try incorporating one vegetarian meal per week, gradually expanding your plant-based options over time.

Sustainable packaging plays a vital, often overlooked, role in reducing environmental harm. Opting for products with minimal packaging whenever feasible is a straightforward step towards sustainability. Choose items in bulk or unpackaged whenever possible, avoiding excessive plastic wrapping. Support companies committed to eco-friendly packaging materials, like recycled paper or biodegradable alternatives. By favoring reusable containers and shopping bags, you are actively participating in curbing plastic waste and contributing to the preservation of natural resources. Consider the entire lifecycle of a product's packaging, from its creation to its disposal, before making purchasing decisions. Small shifts in consumer behavior can accumulate, resulting in a large reduction of waste.

Ethical consumption goes beyond simply buying eco-friendly products; it involves considering the entire supply chain and the social and economic dimensions of food production. Support fair trade initiatives that guarantee fair prices and safe working conditions for farmers and laborers. Choose products from companies that prioritize sustainability and environmental protection across their operations. Consider the origin of your food, ensuring it aligns with your values of ethical and sustainable practices. This can be as simple as verifying certifications or searching for brands committed to ethical sourcing. Researching companies, reading labels, and questioning corporate policies are powerful acts of conscious consumerism, positively influencing the entire food system. In conclusion, by consciously choosing sustainable food practices, we each play an essential role in transforming our food systems into ones that are environmentally responsible, socially equitable, and economically viable for generations to come.

6.1 Reducing Food Waste

Reducing food waste is crucial for a sustainable future. Globally, a significant portion of produced food ends up wasted, contributing to environmental damage, economic losses, and food insecurity. This waste occurs throughout the supply chain, from farm to table, impacting both producers and consumers. Understanding the causes and consequences of food waste is the first step toward effective reduction strategies.

The majority of food waste stems from inefficiencies in the production and distribution networks. Farmers may lose crops due to poor storage or spoilage before reaching markets. Retailers and restaurants contribute significantly by discarding perfectly edible products due to aesthetic standards or overstocking. Consumers, in turn, often purchase more food than they can consume, leading to home spoilage. These various factors all need to be tackled collaboratively.

For producers, implementing improved harvesting and post-harvest techniques is paramount. Investing in better storage facilities and technologies that extend shelf life, such as controlled atmosphere storage, can greatly minimize spoilage. Furthermore, adopting precision agriculture techniques can improve yield and reduce losses associated with unpredictable weather events or pest infestations. Efficient transportation and distribution networks are also essential to prevent spoilage during transit.

Retailers play a critical role in reducing food waste at the consumer level. Implementing clear labeling and "best before" dates can help consumers understand product shelf life and avoid unnecessary waste. Participating in food donation programs to redistribute surplus food to food banks or other charities is a beneficial action to take, actively tackling waste instead of simply discarding. Implementing innovative solutions like dynamic pricing, where prices adjust based on remaining shelf life, can incentivize consumers to purchase products nearing their expiration dates.

Consumers themselves have a significant role to play in reducing food waste. Careful meal planning, accurate grocery shopping, and utilizing leftovers creatively are simple yet impactful changes. Proper food storage, including using appropriate containers and refrigerating perishable items correctly, significantly extends shelf life. Learning to preserve food through methods like canning, freezing, or pickling allows consumers to utilize excess produce before it spoils, creating an environment of reduced waste and resourcefulness.

Composting food scraps is another environmentally friendly solution, diverting organic waste from landfills and creating nutrient-rich soil amendments. This practice transforms waste into a valuable resource, contributing to sustainable gardening and reducing the environmental burden associated with waste disposal.

Implementing effective food waste reduction strategies requires a multi-faceted approach that engages all stakeholders, from producers to consumers. Government regulations and policies, such as extending tax breaks for businesses that implement food waste reduction programs or imposing penalties for excessive waste, can also incentivize action. Public awareness campaigns can empower consumers and educate them about their responsibility to reduce waste. Educating individuals about this crucial component of sustainable living leads to better practices.

Ultimately, reducing food waste is not just an environmental imperative but a social and economic one. By minimizing food loss and waste, we can improve food security, reduce environmental impacts, and create more efficient and sustainable food systems. The combined effort of producers, retailers, consumers, and policymakers is essential to achieving this goal. The changes might seem small individually, but together, they can cause a significant shift in societal behavior. Each action taken can make a substantial difference in the long run, leading to a more sustainable and equitable future. Individual effort combined with broader institutional changes is the key to achieving significant reductions in food waste.

6.2 Local and Seasonal Food

Choosing local and seasonal food is more than just a trendy culinary choice; it's a powerful act of sustainable living. It directly supports your local farmers and reduces the environmental impact associated with long-distance transportation and storage. Think about the miles your food travels before reaching your plate—the carbon emissions from trucks, ships, and airplanes add up significantly. By opting for locally sourced produce, you dramatically lessen your carbon footprint.

The taste difference is remarkable. Fruits and vegetables picked at their peak ripeness and consumed shortly afterward offer a vibrancy of flavor and texture simply unavailable in mass-produced, long-distance transported food. That intense freshness translates to a more satisfying culinary experience. This enhanced flavor profiles makes healthy eating more enjoyable, encouraging you to incorporate more fruits and vegetables into your diet. This benefits not only your palate but your overall health.

Furthermore, purchasing local and seasonal produce strengthens your community. By directly supporting local farmers' markets and farms, you're contributing to a more robust and resilient local economy. This sustains jobs within your community and promotes economic diversification, reducing reliance on large-scale, often environmentally damaging, agricultural operations. You're actively participating in creating a more sustainable food system within your area.

Consider the seasonal availability of produce. Enjoying foods when they're naturally abundant is not only more economical—prices typically reflect peak supply—but also significantly minimizes environmental impact. For example, eating strawberries in summer, grown locally, is far more sustainable than eating strawberries flown in from another hemisphere during winter. This seasonal approach aligns consumption with natural cycles, reducing the need for energy-intensive storage and transportation techniques.

Planning your meals around seasonal availability encourages creativity in the kitchen. It compels you to explore new recipes and ingredients, enriching your culinary skills and expanding your palate. Suddenly, the challenge becomes finding ways to creatively use what's fresh and abundant, leading to more interesting and often healthier meal choices. It's an opportunity to connect more deeply with the food you eat.

Beyond fruits and vegetables, consider local meat, dairy, and even grains. Supporting local farmers who prioritize animal welfare and sustainable practices, such as pasture-raised livestock, is another important step. This choice impacts not only your food choices but also your ethical consumption habits, encouraging the widespread adoption of more environmentally friendly agricultural methods. It benefits both the environment and animal well-being.

However, accessing local and seasonal food requires some effort. Farmers' markets, community-supported agriculture (CSA) programs, and farm stands might not be readily available to everyone, particularly in urban settings. But exploring options—finding local food co-ops, joining online farm delivery services, or even starting a small community garden—can help bridge these gaps. The benefits far outweigh the extra effort involved in discovering and supporting local food systems.

Ultimately, embracing local and seasonal food isn't just about eating better; it's about participating in a more sustainable and just food system. It's about building strong community ties, reducing your environmental footprint, and promoting environmentally and socially responsible agricultural practices. It is a decision that reverberates beyond your dinner plate. Making this shift fosters a closer connection with nature and empowers you to positively influence the future of food production. So, explore your local options, experiment with seasonal recipes, and enjoy the rewarding journey towards a more sustainable lifestyle. The rewards—both culinary and environmental—are immeasurable. By consciously choosing local and seasonal produce, you become an active participant in creating a more sustainable food future. It's about actively participating in a healthier planet, one delicious bite at a time. Consider starting small – perhaps by incorporating one local and seasonal ingredient into your weekly meals, gradually increasing your reliance on locally sourced foods. It's a commitment, not a sudden switch, and the cumulative effect of many small changes is significant.

6.3 Dietary Changes

Shifting our diets towards plant-based options is a powerful step towards a more sustainable food system. Reducing meat consumption, particularly red meat, significantly decreases the environmental footprint of our food choices. Livestock farming, especially intensive practices, contributes substantially to greenhouse gas emissions, deforestation, and water pollution. By incorporating more plant-based proteins like legumes, tofu, tempeh, and nuts into our meals, we lessen our reliance on resource-intensive animal agriculture. This transition isn't about eliminating meat entirely, but rather about consciously reducing our consumption and making informed choices about the sources of our animal products. Choosing sustainably raised, grass-fed meats when consuming meat is an alternative to significantly reducing our impact on the environment. The environmental advantages extend beyond greenhouse gas reduction. Plant-based diets generally require less land and water per calorie produced compared to meat-heavy diets. This reduced demand on land resources helps to preserve biodiversity and minimize habitat destruction, contributing to a healthier ecosystem. Furthermore, the production of plant-based foods often involves fewer pesticides and fertilizers, leading to less water pollution and healthier soil. These positive impacts are not only beneficial for the environment but also contribute to public health by reducing exposure to harmful chemicals. We can consider our water footprint when reducing our meat consumption; it is lower when consuming plant-based foods. This is very important when it comes to water management and protecting our water resources.

Consider the practical aspects of incorporating dietary changes. Transitioning to a more plant-focused diet doesn't require drastic overnight changes. Start by gradually increasing the number of plant-based meals you consume per week. Experiment with new recipes and explore different plant-based protein sources. Many resources are available online and in cookbooks to guide this transition. Join online communities or participate in cooking classes focusing on plant-based cuisine. This will make the transition more enjoyable and sustainable in the long run. Remember that even small changes can make a substantial collective impact.

Beyond individual choices, supporting sustainable agricultural practices through our food choices is crucial. Look for certifications that indicate environmentally friendly farming methods, such as organic labels, fair trade certifications, or certifications specifically addressing water conservation or reduced carbon footprint. These certifications provide assurance that your food choices are aligning with sustainability goals. By actively seeking out and purchasing these products, we incentivize farmers to adopt more sustainable practices, creating a positive feedback loop that benefits both the environment and the farmers themselves. Farmers can take actions in this direction, like using alternative pest control methods instead of synthetic pesticides.

Reducing food waste, a significant contributor to environmental degradation, is another vital component of dietary changes toward sustainability. Plan your meals carefully, store food properly, and utilize leftovers creatively. Composting food scraps reduces landfill waste and creates nutrient-rich soil amendments. Understanding portion sizes and avoiding impulse purchases are important aspects of reducing food waste. By adopting these practices, we minimize the resources wasted in producing food that ultimately ends up discarded. Appreciating the resources used in producing our food and making conscious consumption choices can drive profound changes in our dietary habits. The importance of eating seasonal fruits and vegetables, for example, is immense.

The transition to more sustainable diets should also consider the social and economic dimensions. Supporting local farmers markets not only provides access to fresh, seasonal produce but also strengthens local economies and fosters closer connections between consumers and producers. Purchasing locally grown foods reduces transportation emissions and supports community resilience. Choosing food with minimal packaging also plays a role in reducing waste, and supporting businesses that prioritize sustainable packaging contributes to a more environmentally friendly system. There are also other important aspects such as reducing salt and sugar in our diets.

Moreover, it's crucial to examine our consumption patterns beyond individual choices. Advocating for policies that support sustainable agriculture and food systems is essential. This includes promoting policies that incentivize sustainable farming practices, reduce food waste, and support research and development in sustainable food production. Such actions encourage systemic change, going beyond individual actions to foster a broader culture of sustainability. We can support organizations advocating for such policy changes, using our voices to express our concerns and our commitment to a more sustainable future for everyone. These actions, both personal and collective, are essential to ensure the long-term health of our planet. The changes required are significant; however, every action, no matter how small, is a step in the right direction.

6.4 Sustainable Packaging

Sustainable packaging is critical for a truly sustainable food system. It's not just about the materials used; it encompasses the entire lifecycle, from sourcing raw materials to end-of-life management. We must move beyond simply replacing plastic with another material; the goal is to minimize environmental impact at every stage. This demands innovative solutions that consider the whole picture.

Thinking about the environmental footprint of packaging requires a shift in perspective. We need to consider not just the material itself, but also the energy required for production, transportation, and disposal. A seemingly "green" material might have a surprisingly large carbon footprint if its production is energy-intensive or relies on unsustainable harvesting practices. For example, some bioplastics boast compostability but their production might require significant land use or water consumption. A complete life cycle assessment is vital.

Choosing the right materials is only one piece of the puzzle. Packaging design plays a significant role. Minimizing material usage through efficient designs directly reduces resource consumption and waste. Lightweight packaging reduces transportation costs and fuel usage, while clever designs can enhance product protection, extending shelf life and reducing food waste. This interconnectivity is key. For example, a sturdy yet lightweight cardboard box, sourced sustainably, is better than a heavier, less efficient alternative, regardless of its recycled content.

The concept of recyclability is equally complex. While recycling is often touted as a solution, it's crucial to understand its limitations.

Not all materials are easily recyclable, and the recycling infrastructure varies greatly across regions. Packaging should be designed for easy recyclability, using materials and designs compatible with existing infrastructure. Furthermore, promoting a circular economy approach, where materials are recovered and reused, should be the primary objective. The ultimate goal should always be to keep materials in the loop.

Consumers also have a vital role to play. Educating consumers about the importance of sustainable packaging and providing clear and readily available recycling information empowers them to make informed choices. Furthermore, encouraging consumers to participate in recycling programs and to support companies committed to sustainable packaging practices can create a powerful market incentive. This collective action creates a more profound impact.

Composting, especially for biodegradable packaging, represents a powerful method for waste reduction and soil enrichment. However, the success of composting depends on the availability of effective composting systems and the education of consumers on proper composting techniques. Furthermore, the design of biodegradable packaging must be consistent with the capacity of local composting systems to ensure efficient breakdown. Without proper management, even biodegradable packaging could become an environmental burden.

Policy and regulation are indispensable to advancing sustainable packaging. Governments can incentivize the use of sustainable materials through tax breaks or subsidies, while regulations can set minimum standards for packaging design and recyclability. Furthermore, governments can invest in improved recycling infrastructure and enforce extended producer responsibility schemes, holding manufacturers accountable for the end-of-life management of their packaging. Effective regulation is a crucial driver of change.

Technological innovation promises further advancements in sustainable packaging. Researchers are developing new bio-based materials with enhanced properties, improving biodegradability and compostability. Innovative design techniques utilizing less material and advancements in recycling technologies are also pushing the boundaries of sustainable packaging solutions. This continuous innovation promises a more sustainable future.

In conclusion, sustainable packaging is not a standalone issue; it is intrinsically linked to reducing food waste, optimizing agriculture, and creating a circular economy. By integrating sustainable packaging strategies with broader sustainability goals, we can significantly contribute to a more environmentally responsible and resilient food system.

6.5 Ethical Consumption

Ethical consumption isn't just about buying organic; it's about understanding the entire lifecycle of a product, from its origin to its eventual disposal. It requires us to be mindful consumers, questioning where our food comes from and the impact of our choices on farmers, workers, and the environment. This demands a shift in our thinking, moving beyond mere price and convenience to consider the broader social and environmental consequences.

Consider the coffee you drink each morning. Do you know where the beans are grown? Were the farmers who harvested them paid a fair wage? Were environmentally sound practices employed during cultivation and processing? More likely than not, the answer to at least one of these questions is no. Many coffee farmers struggle to make a living wage, and the environmental costs of coffee production, from deforestation to pesticide use, are substantial. Choosing ethically sourced coffee, however, supports sustainable farming practices and fairer labor conditions.

Ethical consumption extends beyond individual products; it includes our overall dietary choices. The meat industry, for example, is a significant contributor to greenhouse gas emissions, deforestation, and water pollution. A reduction in meat consumption, especially red meat, can dramatically decrease one's environmental footprint. This isn't necessarily about becoming a vegetarian or vegan, but rather about making conscious choices—perhaps opting for plant-based meals more often or choosing meat from sustainably raised animals.

Furthermore, reducing food waste is a crucial aspect of ethical consumption. A substantial portion of food produced globally is wasted, contributing to resource depletion and environmental damage. By planning meals, storing food properly, and composting food scraps, we can minimize our contribution to this problem. This simple act significantly reduces our impact.

Moreover, supporting local and seasonal produce reduces transportation costs and emissions associated with food distribution. Farmers' markets and community-supported agriculture (CSA) programs provide direct access to locally grown food, often produced using more sustainable methods. Knowing your farmer and the origin of your food strengthens the connection between consumer and producer, fostering a greater sense of responsibility and community.

The packaging of our food also presents ethical considerations. Excessive packaging contributes to plastic pollution and waste. Choosing products with minimal packaging or opting for reusable containers whenever possible reduces our environmental impact. Moreover, supporting companies that prioritize sustainable packaging practices demonstrates our commitment to environmental responsibility and pushes for industry-wide change.

However, ethical consumption is not always easy. Ethically sourced products often come with a higher price tag, and access to these products might be limited depending on location and availability. This creates a challenge, particularly for those with limited financial resources. It is crucial to acknowledge this disparity and work towards equitable access to sustainable and ethical products for all consumers.

It's also vital to remember that ethical consumption is not about perfection. It's a journey, a continuous process of learning and improvement. Start small, focus on making incremental changes, and celebrate every step forward. The impact of even small changes can be significant when multiplied across a large number of people. Remember that making informed choices requires research. It might involve reading product labels, investigating company practices, and looking for certifications that verify ethical and sustainable production. These certifications, such as Fair Trade or organic labels, can serve as helpful guidelines, although it's always advisable to verify their authenticity and comprehensive nature through independent investigation.

Ultimately, ethical consumption requires a holistic approach. It involves considering the social, economic, and environmental implications of our consumption habits and acting in a way that promotes fairness, sustainability, and well-being for all. It demands active engagement, continuous learning, and a willingness to challenge our consumption patterns and advocate for change. It's about making mindful choices, supporting responsible businesses, and reducing our overall impact on the planet. It's a collaborative effort, not an individual burden. By combining personal responsibility with collective action, we can transform our consumption habits and create a more just and sustainable food system. It's about actively shaping a better future, a future where economic prosperity, social equity, and environmental health are no longer mutually exclusive concepts but mutually reinforcing pillars of a thriving society. Our choices matter. Our collective choices matter even more. Let's make them count.

7 Policy and Future Directions

Government regulations are essential for driving the transition to sustainable agriculture. Incentive-based policies, such as subsidies for organic farming or carbon sequestration projects, can significantly impact farmer adoption of sustainable practices. Conversely, regulations that restrict the use of harmful pesticides or fertilizers can protect the environment and public health. Effective policies need to consider regional variations in agricultural practices and ecological conditions; a "one-size-fits-all" approach rarely works. Careful consideration must also be given to the economic impact on farmers, ensuring a just transition that doesn't leave them financially disadvantaged. Stronger environmental protection laws, including stricter regulations on water and air pollution stemming from agricultural activities, are vital for long-term sustainability. This also necessitates robust monitoring and enforcement mechanisms to ensure compliance and accountability. International collaborations are crucial; global agreements on trade and environmental standards can help create a level playing field and promote responsible agricultural practices worldwide. Furthermore, policies should actively support research and development in sustainable agricultural technologies.

Technological advancements offer tremendous potential for enhancing sustainable agriculture. Precision agriculture, using GPS-guided machinery and sensor technology, optimizes resource use, minimizing waste and maximizing yield. Drones can be used for crop monitoring, early pest detection, and targeted application of fertilizers and pesticides, reducing overall input use. Data analytics and machine learning can help farmers make informed decisions based on real-time data, leading to improved efficiency and reduced environmental impact. Developments in biotechnology, such as disease-resistant crops and nitrogen-fixing microbes, can further enhance sustainability by reducing reliance on chemical inputs. Renewable energy technologies, such as solar and wind power, can reduce agriculture's carbon footprint by powering farm operations. Improved water harvesting and irrigation techniques, like drip irrigation systems, increase water-use efficiency. Vertical farming and hydroponics, while currently expensive, hold potential for increased food production in urban areas, reducing transportation distances and associated emissions. Investment in research and development of these technologies, combined with appropriate educational programs to train farmers in their use, is paramount.

The future of sustainable agriculture hinges on a multifaceted approach that integrates policy and technological innovation. Policies must incentivize adoption of environmentally friendly practices and provide support for farmers during the transition. Technological advances provide the tools for increased efficiency and reduced environmental impact; however, their implementation requires careful planning and consideration for equitable access. Furthermore, building resilient food systems requires collaboration across sectors – farmers, researchers, policymakers, and consumers need to work together to create a future where food production is both sustainable and secure. Education and awareness-raising campaigns can empower consumers to make responsible choices, supporting sustainable agricultural practices through their purchasing decisions. International cooperation is also critical, as climate change and global food security are interconnected challenges demanding global solutions. Stronger regulatory frameworks combined with technological innovation can pave the way for truly sustainable food production systems capable of feeding a growing global population without compromising environmental integrity. Ultimately, success depends upon a systemic transformation, driven by a shared understanding of the urgency and the long-term benefits of embracing sustainable agricultural practices. By fostering collaboration, implementing effective policies, and harnessing the power of technological innovation, we can secure a future where food production is both sustainable and secure for generations to come. The challenges are significant, but the rewards—a healthy planet and a food-secure future—are immeasurable. A coordinated global effort is not just desirable but essential for achieving these ambitious goals.

7.1 Government Regulations

Government regulations play a crucial role in steering the agricultural sector towards sustainability. Effective policies can incentivize farmers to adopt environmentally friendly practices, while poorly designed regulations can hinder progress or even create unintended negative consequences. The complexity lies in balancing economic viability for farmers with the pressing need for environmental protection and food security. A well-crafted regulatory framework should not be a burden, but rather a catalyst for innovation and responsible farming.

Many countries employ a combination of approaches, including direct financial incentives such as subsidies for organic farming or water-efficient irrigation systems. These subsidies can significantly lower the initial investment costs associated with adopting sustainable practices, making them more accessible to a wider range of farmers. However, careful consideration must be given to ensure these subsidies are targeted effectively, avoiding perverse incentives that might inadvertently support unsustainable practices. For example, poorly designed subsidies could encourage overproduction of certain crops, leading to market imbalances and waste.

Beyond direct financial support, governments can implement regulations that set standards for pesticide use, fertilizer application, and water management. These regulations, often backed by penalties for non-compliance, provide a crucial framework for responsible resource use. Examples include limits on the use of specific pesticides known to harm beneficial insects or contaminate water sources. Similarly, regulations might mandate the use of specific soil conservation techniques or set limits on water extraction from aquifers. However, the effectiveness of such regulations relies heavily on robust monitoring and enforcement mechanisms. Without effective oversight, regulations remain mere suggestions, failing to achieve their intended impact.

The development of effective regulations requires a collaborative approach involving policymakers, scientists, farmers, and consumer advocacy groups. Open dialogue and transparent data sharing are paramount. Policies should be evidence-based, grounded in sound scientific research and understanding of local ecological and socioeconomic conditions. Ignoring the specific contexts of different agricultural regions and farmer livelihoods could lead to policies that are ineffective or even counterproductive. Furthermore, it's crucial to involve farmers in the development of regulations; their practical knowledge and perspectives are essential to ensure that policies are both effective and feasible to implement.

One critical aspect often overlooked is the enforcement of existing regulations. Strong enforcement is crucial for maintaining compliance and preventing environmentally damaging practices. This requires sufficient resources allocated to monitoring and inspection programs. Moreover, penalties for non-compliance must be substantial enough to deter violations but also fair and transparent to avoid driving farmers out of business. Finding the appropriate balance between environmental protection and economic sustainability remains a significant challenge in regulatory design.

Furthermore, the impact of government policies extends beyond individual farms to encompass broader supply chains. Regulations influencing sustainable packaging, food waste reduction, and consumer choices all play a significant role in the overall sustainability of the agricultural system. Governments can incentivize businesses to adopt environmentally friendly packaging materials and reduce food waste along the supply chain, from farms to retailers and consumers. Likewise, public awareness campaigns can empower consumers to make more informed and sustainable food choices. This holistic approach, considering the entire food system, is necessary for creating truly lasting change.

The development and implementation of effective government regulations require ongoing adaptation and refinement. Regular evaluation and monitoring are necessary to assess the impact of policies and make necessary adjustments. The agricultural landscape is constantly evolving, with new technologies and challenges constantly emerging. Regulations must evolve alongside these developments to maintain their relevance and effectiveness. Flexibility and adaptability are key features of a successful regulatory framework for sustainable agriculture. Continuous learning, feedback mechanisms, and iterative adjustments are vital to ensure that policies remain relevant and effective in promoting a genuinely sustainable agricultural system.

Ultimately, effective government regulations are essential for driving the transition to sustainable agriculture. However, it is not simply about creating new regulations; it is about building a coherent and consistent policy framework. This framework must be rooted in scientific understanding, economically viable for farmers, and adaptable to changing circumstances. By combining targeted financial incentives with strong regulatory standards and active engagement with stakeholders, governments can play a powerful role in ensuring the long-term health of our planet and food security for generations to come. The success of these efforts relies heavily on collaboration, transparency, and a commitment to continuous improvement.

7.2 Technological Advancements

Precision agriculture technologies are revolutionizing farming, offering unprecedented levels of control and efficiency. GPS-guided machinery allows for pinpoint application of fertilizers, pesticides, and water, minimizing waste and maximizing resource utilization. This targeted approach reduces environmental impact significantly, cutting down on chemical runoff and unnecessary energy expenditure. Furthermore, remote sensing via drones and satellites provides real-time data on crop health, soil conditions, and irrigation needs, enabling farmers to make informed decisions promptly and optimize yields. Data analysis tools process this information, identifying areas needing attention and suggesting timely interventions. This sophisticated system empowers farmers to act proactively rather than reactively, boosting efficiency and sustainability. The development of drought-resistant and pest-resistant crops through genetic engineering and biotechnology offers a powerful tool in combating climate change and reducing reliance on chemical interventions. Modifying crops to withstand harsh conditions decreases the need for extensive irrigation and pesticide application, preserving water resources and protecting biodiversity. These advancements are vital in regions facing water scarcity or extreme weather patterns, enhancing food security in vulnerable areas. However, careful consideration of ethical implications and potential unintended consequences is crucial for responsible implementation. Rigorous testing and transparent regulatory processes are essential to ensure the safe and sustainable deployment of these technologies.

Vertical farming, hydroponics, and aeroponics are innovative techniques that maximize food production in limited spaces. These methods utilize controlled environments to optimize growing conditions, drastically reducing water and land requirements compared to traditional farming. Vertical farms, in particular, can be located in urban areas, minimizing transportation distances and reducing the carbon footprint associated with food distribution. Hydroponics and aeroponics further enhance efficiency by providing nutrients directly to plant roots, maximizing resource absorption. Although the initial investment in these technologies can be high, the long-term benefits in terms of resource conservation and increased yields are substantial, making them increasingly attractive options for sustainable food production.

Robotics is transforming agricultural practices, automating tasks previously requiring manual labor. Robots can perform tasks such as planting, weeding, harvesting, and spraying, increasing efficiency and reducing labor costs. Autonomous tractors and harvesters, guided by GPS and sensors, work tirelessly, precisely executing farming operations day and night. This technology not only boosts productivity but also improves worker safety by removing them from hazardous environments. The integration of AI and machine learning algorithms further enhances robotic capabilities, enabling them to adapt to changing conditions and optimize their performance. While widespread adoption requires considerable investment in infrastructure and training, the potential for improving efficiency and sustainability is immense.

Renewable energy sources, such as solar and wind power, are increasingly powering agricultural operations, reducing reliance on fossil fuels. Solar panels can provide electricity for irrigation pumps, lighting, and processing equipment. Wind turbines can generate electricity for larger farms. This shift towards renewable energy sources not only reduces greenhouse gas emissions but also decreases energy costs in the long run, contributing to the economic viability of sustainable agricultural practices. Furthermore, integrating renewable energy with other smart farming technologies can create a truly sustainable and self-sufficient farming system. Investing in renewable energy infrastructure is crucial for minimizing the environmental impact of agriculture while enhancing its long-term sustainability and resilience.

Big data analytics and machine learning are playing a crucial role in optimizing agricultural practices. By analyzing vast amounts of data from various sources, farmers can gain insights into crop performance, soil conditions, and weather patterns, allowing them to make data-driven decisions that improve yields, reduce resource use, and mitigate risks. Precision agriculture technologies generate a wealth of data, which, when effectively analyzed, can lead to significant improvements in sustainability and profitability. Machine learning algorithms can identify patterns and trends in data that might be missed by human analysis, leading to more precise predictions and informed decisions. Investing in data analytics capabilities is essential for harnessing the full potential of modern agricultural technologies.